I0409424

Space Adventures with Quantum Quark

Tihirou Muctarr Nicol

Table of Contents

Copyright © 2023 by Tihirou Muctarr Nicol

All rights reserved. No part of this work may be reproduced, distributed, or transmitted in any form or by any means, including photocopying, recording, or other electronic or mechanical methods, without the prior written permission of the copyright holder, except for brief quotations in critical reviews and other noncommercial uses permitted by copyright law.

Limit of Liability/Disclaimer of Warranty: The author and publisher have made every effort to ensure the accuracy and completeness of the contents of this work but make no warranties, expressed or implied, regarding errors or omissions. The information provided in this work is sold with the understanding that the author and publisher are not engaged in rendering legal, financial, or other professional advice. If professional assistance is required, the services of a competent professional should be sought.

For permission requests, please contact the publisher at the following address:

10 Dockyard Road

Kissy, Freetown, Sierra Leone

Tihirounicol@gmail.com

Tyronenicol.com

This work is a piece of fiction. Names, characters, places, and incidents are either the product of the author's imagination or are used fictitiously. Any resemblance to actual persons, living or dead, events, or locales is entirely coincidental.

Printed in the United States of America

Dedication

I wish to dedicate this book and extend a heartfelt "Thank You" to every purchaser and potential supporter who, at some point, responded with a "No."

I JUST MADE A SALE!

You're my new customer, and I want to express my heartfelt thanks for choosing to do business with me. Your support means the world to me, and I value your trust. In my work, which includes consulting, volunteering, and writing, I have three key goals:

- To make a positive impact on people's lives.
- To build strong, lasting relationships.
- To have a blast while doing it!

When I wrote this book, my main aim was to make it incredibly useful to you. I hoped it would be so helpful that you'd happily recommend it to at least ten of your co-workers, friends, and family members. I'd love to hear from you if I've succeeded in reaching that goal.

Because of you, and all my amazing customers, I get to do what I'm passionate about – selling, writing, and teaching.

Thank you so much!

Introduction

In the vast cosmic expanse, where stars twinkle like diamonds and galaxies stretch beyond imagination, there exists a realm of science and wonder that is as captivating as the universe itself - the world of quantum physics. Join us on an extraordinary journey through space and time with our endearing guide, Quantum Quark, as we embark on a series of educational space adventures designed especially for curious young minds.

In "Space Adventures with Quantum Quark," we will unravel the enigmatic and fascinating concepts of quantum physics in a way that is engaging, entertaining, and educational. Quantum Quark, a charming and curious particle, will lead us through the realms of the very small, where the laws of classical physics no longer apply, and the bizarre and mind-bending rules of quantum mechanics take center stage.

Throughout this series of books, we will explore the mysteries of quantum entanglement, the uncertainty principle, the wave-particle duality, and much more. Each chapter will take you on a thrilling journey through space, time, and the quantum realm, all while providing a solid foundation in the fundamental principles of physics.

Prepare to be amazed as we dive headfirst into black holes, traverse wormholes, and even venture into parallel universes, all guided by the delightful Quantum Quark. This series is not just a voyage through space but a journey into the heart of science, where questions lead to discoveries, and discoveries lead to more questions.

So, fasten your seatbelts and get ready for an adventure of cosmic proportions. It's time to unlock the secrets of the universe with "Space Adventures with Quantum Quark."

Chapter 1: The Mystery of Quantum Physics

Hey there, fellow space adventurers! Buckle up your cosmic seatbelts because we're about to dive headfirst into a mind-blowing realm - the quirky, quantum world. And guess what? Our tour guide for this wild journey is none other than the one and only Quantum Quark!

You might be wondering, "What's the big deal with quantum physics anyway?" Well, hold onto your helmets, because this stuff is the real cosmic deal.

Picture this: you're in a sci-fi movie, and your spaceship is zooming through the universe. But instead of stars and planets, you're surrounded by things that don't quite make sense. Imagine a coin flipping through the air, and in some crazy twist of fate, it's both heads AND tails at the same time. Mind-boggling, right? That's the quantum world for you.

In our everyday world, we're used to things being one way or the other. Cats are either in boxes or out of boxes, right? But in the quantum realm, things aren't so straightforward. Tiny particles, like electrons and photons, have a knack for being in multiple places at once. It's like they're the ultimate multitaskers of the universe.

And here's where it gets even more bonkers. Quantum Quark is about to spill the quantum beans on something called "superposition." This is like particles showing off their acrobatic skills. They can exist in multiple states simultaneously. It's as if they're saying, "Why be just one thing when I can be all the things at once?"

Now, let's talk about quantum entanglement. Imagine you and your best space buddy have a magical connection. You both go your separate ways, but no matter how far apart you are, if one of you does a somersault, the other one instantly somersaults too. That's entanglement in a nutshell - particles becoming quantum BFFs, no matter the distance.

Our fearless guide, Quantum Quark, loves a good riddle. So here's one for you: What's both a particle and a wave, but only when you're not looking? The answer: Light! That's right; light can be a particle (like tiny bits of stuff) and a wave (like ripples in a pond). This switcheroo between roles is called "wave-particle duality," and it's another mind-blowing feature of the quantum world.

Now, let's meet Heisenberg's Uncertainty Principle - the cosmic rule that keeps us all on our toes. Imagine trying to measure a particle's position and speed at the same time. Well, Heisenberg says, "Hold on, folks, you can't know both precisely!" It's like trying to catch a firefly in the dark - you can either know where it is or how fast it's zipping around, but not both. This uncertainty is built into the quantum fabric of the universe.

So, what's the takeaway from this whirlwind tour of quantum physics? First, quantum physics is like the universe's hidden treasure chest of weirdness. It challenges everything we think we know about the cosmos. Second, Quantum Quark is here to be your trusty guide through the twists and turns of this quantum rollercoaster. Together, we'll tackle the big questions and mind-bending mysteries that make the quantum world so fascinating.

As we journey deeper into the quantum cosmos in the chapters ahead, get ready to explore black holes, time travel, and even parallel universes. With Quantum Quark leading the way, there's no adventure too wild, no concept too mind-boggling. So, fasten your seatbelts, my fellow space adventurers, and let's keep riding this cosmic wave into the heart of quantum physics. It's going to be a quantum-tastic journey!

Chapter 2: Quantum Quark's Amazing Journey

Alrighty, space adventurers, it's time for Chapter 2 of our quantum-tastic adventure with none other than Quantum Quark! Now, if you thought Chapter 1 was a mind-bender, you better hold onto your space helmets because we're about to blast off into Quantum Quark's Amazing Journey through the quantum cosmos.

So, picture this: Quantum Quark is this super tiny, ultra-curious particle. Think of it as that little spark of curiosity inside you, but with a mischievous grin and the power to zip around the universe. Quantum Quark might be small, but it's got a heart as big as a black hole, and it's on a mission to show us just how wild and wacky the quantum world can be.

Now, in our last chapter, we talked about superposition, where particles like to be in multiple places at once. Well, Quantum Quark isn't just any particle; it's the master of superposition. It can be here, there, and everywhere all at the same time. Imagine having a superpower like that! Quantum Quark uses this incredible ability to take us on a journey that defies the laws of classical physics.

As we tag along with Quantum Quark, we find ourselves in a place where the rules of space and time don't quite work the way we're used to. It's like stepping through a cosmic wormhole into a realm where reality itself is up for grabs. Picture this: you're in a room with a light bulb, and you turn it on. Simple, right? But in the quantum world, things get a tad... weird.

Quantum Quark introduces us to a thought experiment that'll make your head spin - Schrödinger's Cat. Imagine a cat in a box with a deadly device that might or might not go off. Until you open the box and look inside, the cat is both alive and... not so alive, all at the same time. It's like saying "yes" and "no" to ice cream simultaneously! This is how the quantum world operates, with possibilities existing in a mind-boggling superposition until observed.

Now, let's talk teleportation. Yep, you heard right, teleportation! In the quantum world, particles can magically "teleport" from one place to another without moving through the space in between. It's like saying, "Beam me up, Quantum Quark!" and poof you're on the other side of the universe. This teleportation trick is thanks to another quantum phenomenon called "quantum tunneling," where particles sneak through barriers, they shouldn't be able to cross. It's the ultimate cosmic shortcut.

But our quantum journey doesn't stop there. Quantum Quark takes us on a cosmic rollercoaster ride through something called "quantum interference." Imagine throwing two pebbles into a pond, and the ripples from each pebble collide and create a wild, wavy dance. Well, in the quantum world, particles do the same thing, creating mesmerizing patterns of interference. It's like the universe is a gigantic cosmic disco, and particles are grooving to their own quantum beats.

Now, I know this all sounds like science fiction, but it's science FACT! And Quantum Quark is here to prove it. With each twist and turn of our journey, Quantum Quark shows us that the quantum world isn't just strange; it's a cosmic playground of infinite possibilities.

So, what's the big lesson from Quantum Quark's Amazing Journey? It's that the universe is way more mysterious and magical than we ever imagined. The quantum world might seem like science fiction, but it's science doing its thing in the most extraordinary way possible.

As we venture further into our quantum quest, remember that Quantum Quark is our trusty guide through this quantum wonderland. Together, we'll explore more cosmic conundrums, tackle paradoxes, and uncover the quantum secrets that make the universe tick. So, space adventurers, fasten your seatbelts and get ready for the ride of a lifetime because Quantum Quark's Amazing Journey is just beginning, and it's going to be a quantum blast!

Chapter 3: Exploring the Quantum World

Hey there, cosmic explorers! Welcome back to our wild journey through the quantum cosmos with Quantum Quark. In Chapter 2, we learned that the quantum world is a place of superpositions, teleportation, and cosmic dance-offs. Now, get ready to blast off again as we dive into Chapter 3: Exploring the Quantum World.

Picture this: You're on a spaceship, and Quantum Quark is your trusty co-pilot, guiding you through the quantum wilderness. We're about to explore the nooks and crannies of the quantum world and trust me, it's like nothing you've ever seen.

So, first things first, let's talk about the quantum zoo. Imagine a place filled with all sorts of quantum creatures, like particles and atoms. But here's the kicker: these critters don't always behave themselves. They're like intergalactic pranksters playing hide-and-seek with the laws of physics.

One of our favorite quantum creatures is the electron. It's a tiny, zippy thing that orbits the nucleus of an atom. But here's the twist: electrons don't follow neat, predictable paths like cars on a highway. Nope, they're more like cosmic acrobats, leaping from one energy level to another in the blink of an eye. It's as if they're saying, "Watch me defy gravity, and oh, by the way, I'm everywhere and nowhere at once!"

Quantum Quark loves to take us on a journey through the double-slit experiment. It's like a quantum magic trick. Imagine shining light or firing particles at a barrier with two slits. You'd expect them to go through one slit or the other, right? But in the quantum world, particles pull a fast one. They go through both slits at the same time, creating a cosmic interference pattern that's as baffling as it is beautiful.

Now, let's talk about something that'll tickle your taste buds - quantum states. Imagine you have a cosmic menu of ice cream flavors, but instead of choosing just one, you can savor them all at once. That's what particles do in the quantum world. They exist in multiple states simultaneously, like a flavor explosion in the universe. Quantum Quark is always up for a quantum ice cream party!

But hold onto your helmets because it's time to tackle the quantum cat again! Schrödinger's Cat is back, and this time, it's bringing its pals, the "quantum coins" with it. Imagine you have a magical coin that can be both heads and tails at the same time. It's like playing a never-ending game of cosmic coin toss, where the outcome is a mystery until you look.

Now, here's where it gets really fun. Quantum Quark introduces us to something called "quantum entanglement." Remember those particles that become BFFs and share secrets across the universe, no matter how far apart they are? Well, quantum entanglement is like the ultimate cosmic phone call between particles. They can communicate faster than the speed of light, and when one of them does a quantum jig, the other one instantly joins the dance. It's like they're connected by an invisible cosmic string.

But what's the big takeaway from Exploring the Quantum World? It's that the quantum universe is a place of wonder, where particles defy the laws of classical physics and do their own quantum dance. It's a cosmic carnival of infinite possibilities, and Quantum Quark is our guide through this magical realm.

As we venture further into the quantum cosmos in the chapters ahead, get ready to explore even more mind-bending concepts. We'll dive into black holes, travel through time, and peek into parallel universes. With Quantum Quark leading the way, there's no quantum conundrum too puzzling, no quantum mystery too mysterious. So, fellow cosmic explorers, fasten your seatbelts and get ready for the next leg of our quantum quest because the quantum adventure is just beginning, and it's going to be a wild ride through the cosmos!

Chapter 4: Quantum Quark and the Quantum Leap

Hey there, cosmic thrill-seekers! It's time for the next exciting chapter of our quantum adventure with Quantum Quark. So far, we've explored superpositions, quantum creatures, and even took a spin with Schrödinger's Cat. But hold onto your helmets because Chapter 4 is all about "Quantum Quark and the Quantum Leap."

Picture this: Quantum Quark is our trusty guide, and we're on the edge of a cosmic cliff. But this cliff isn't just any old cliff; it's a quantum cliff. And guess what? Quantum Quark is about to take the most daring leap of its tiny life.

Now, remember those electrons we talked about in the last chapter? They're like the cosmic acrobats of the quantum world, always jumping from one energy level to another. Well, Quantum Quark is about to teach us how they do it. It's like learning the secrets of the universe's greatest trampoline act.

Imagine an electron in its cozy little orbit around an atom's nucleus. Everything's calm and steady, like a cosmic ballet. But then, out of nowhere, Quantum Quark gives that electron a nudge - a quantum nudge, to be precise.

This nudge sends the electron soaring to a higher energy level, but here's the kicker: it doesn't follow a nice, predictable path to get there. Nope, it just vanishes from its original orbit and reappears in the higher one, as if it teleported through space. It's like saying, "Hey, lower energy level, I've got places to be - see ya!" This incredible leap is what we call a "quantum leap."

Now, you might be wondering, "How can an electron just disappear and reappear like that?" Well, welcome to the wacky world of quantum physics! In this realm, particles don't follow the same rules as everyday objects. They can exist in multiple places at once and jump between energy levels without going through the space in between.

Quantum Quark loves to call this phenomenon "quantum magic," and it's easy to see why. It's like particles have their own secret teleportation devices hidden in their pockets. But there's more to this quantum leap than meets the eye.

Quantum Quark explains that electrons can only leap to specific energy levels, like stepping stones in a cosmic pond. They can't just jump anywhere; they have to follow the quantum rules. And when they make these leaps, they release or absorb packets of energy called "quanta." It's like the universe's way of saying, "You've done it! Here's your cosmic reward!"

Now, let's talk about something called "quantum tunneling." Imagine you're on one side of a massive mountain, and you want to get to the other side, but there's no way around it. In the classical world, you'd be stuck. But in the quantum world, particles are like the ultimate cosmic mountain climbers. They can magically tunnel through barriers they shouldn't be able to cross. It's like saying, "I don't need a pickaxe; I'll just quantum-tunnel my way through!"

Quantum tunneling is responsible for some mind-blowing phenomena, like nuclear fusion in stars and the operation of transistors in your gadgets. It's proof that the quantum world isn't just strange; it's incredibly powerful and influential in our everyday lives.

So, what's the big lesson from Quantum Quark and the Quantum Leap? It's that the quantum world is full of surprises and cosmic acrobatics. Particles like electrons don't follow the same rules as our everyday experiences, and they can do things that seem impossible in the classical world.

As we journey deeper into the quantum cosmos in the chapters ahead, get ready to explore even more mind-bending concepts. Quantum Quark will lead us through the twists and turns of the quantum universe, where the rules of reality are constantly being rewritten. So, space adventurers, fasten your seatbelts and prepare for the next quantum leap because the quantum adventure is far from over, and it's going to be a cosmic thrill ride like no other.

Chapter 5: Quantum Quark's Quantum Friends

Hey there, fellow cosmic explorers! Welcome back to our quantum adventure with Quantum Quark. We've been on quite the ride through the quantum universe, from superpositions to quantum leaps. Now, in Chapter 5, we're about to meet some fascinating characters in Quantum Quark's Quantum Friends.

So, picture this: Quantum Quark isn't just a lone particle on this wild journey; it's got a whole entourage of quantum buddies. These pals are like the Avengers of the quantum world, each with its own unique superpower. Let's dive in and get to know a few of them.

First up, meet the Electron Duo - Spin and Charge. These dynamic particles are the life of the quantum party. Spin is like the cosmic dance move of electrons, and it can be either "up" or "down." Imagine a particle twirling like a tiny top while also being in two places at once. That's Spin for you. Then there's Charge, which is like the electron's superhero name tag. It's either positive or negative, and it's how particles play nice and keep their cosmic dance organized.

But here's where things get really cool. Spin and Charge are like best buddies that love to tango. They have this amazing dance routine called "spin-orbit coupling." It's like they're doing the quantum waltz, and their dance creates some mind-bending effects in the quantum world. This duo is responsible for the magic of magnetic fields and the behavior of electrons in atoms. They're like the ultimate quantum dance partners!

Now, let's meet the Photonic Twins - Particle and Wave. These two are the ultimate quantum shape-shifters. Particle is like the tiny bit of stuff that makes up light, and it's all about being in one place at a time. But then, there's Wave, which is like the ripples on a cosmic pond. It's all about spreading out and being in many places at once. Sounds contradictory, right?

Well, here's the twist - Photonic Twins can switch between these two forms whenever they want. It's like saying, "Hey, I feel like being a particle today, but tomorrow, I'll be a wave!" This phenomenon is what we call "wave-particle duality," and it's one of the quantum world's greatest mysteries.

But the Photonic Twins aren't just shape-shifters; they're also cosmic messengers. They carry information as light waves, making things like lasers and fiber-optic communication possible. So, the next time you're streaming your favorite space adventure movie, thank Particle and Wave for their quantum message delivery!

Now, let's talk about the Quark Triplets - Up, Down, and Strange. These three are like the rockstars of the quantum world. Quarks are the tiny building blocks of protons and neutrons, which, in turn, make up the nuclei of atoms. Up and Down quarks are like the cosmic glue that holds everything together in the atomic nucleus. They're the cosmic cement!

But then there's Strange quark, the rebel of the trio. Strange quark is like the cosmic party crasher. It brings unpredictability and strangeness to the quantum gathering. When it shows up, it transforms into other particles, creating a quantum carnival of particle transformations.

These quirky quarks are proof that the quantum world isn't just about particles; it's about the incredible diversity of particles and their unique personalities. They make up the cosmic cast of characters in the quantum universe, each playing a vital role in the grand quantum story.

So, what's the big takeaway from meeting Quantum Quark's Quantum Friends? It's that the quantum world is a bustling cosmic community, filled with particles that have their own superpowers and quirks. They dance, shift shapes, and bring order to the quantum chaos in their own special ways.

As we venture deeper into the quantum cosmos in the chapters ahead, remember that Quantum Quark and its quantum buddies are here to show us that the quantum world is not just strange; it's a vibrant and diverse place where particles party, waltz, and tango through the cosmic dance of existence. So, fellow cosmic explorers, fasten your seatbelts because the quantum adventure is far from over, and Quantum Quark's Quantum Friends are just getting started with their quantum shenanigans!

Chapter 6: The Quantum Quark Paradox

Hey there, fellow quantum enthusiasts! Are you ready for the next mind-bending chapter of our quantum adventure with Quantum Quark? Well, grab your cosmic popcorn because Chapter 6 is all about the "Quantum Quark Paradox." Trust me, it's a puzzle that'll make your brain do somersaults in the vast quantum playground.

Picture this: Quantum Quark is in a playful mood, and it's about to lead us into the heart of a paradox - a cosmic conundrum that'll leave you scratching your space helmet.

Now, you've probably heard that in the quantum world, particles can be in multiple places at once, thanks to a phenomenon called superposition. But here's the kicker: the moment you observe a particle, it suddenly decides to be in just one place, as if it's camera-shy and doesn't want you to catch it being in two places at the same time.

This phenomenon is known as "wave function collapse," and it's at the heart of the Quantum Quark Paradox. It's like particles are playing hide-and-seek with the laws of classical physics. As soon as you peek, they say, "Aha! You found me! I was just here all along," and they act as if they were never in that tricky superposition.

But here's where the paradox gets paradoxical. Quantum Quark asks, "What if we don't look? What if we leave the particle alone and let it do its quantum thing?" Well, in that case, the particle remains in a superposition of states, existing in multiple places at once, like a cosmic multitasker.

So, you might be thinking, "What's the big deal? Let the particle do its superposition dance, right?" But here's the cosmic twist: once you do decide to look, the particle instantly snaps out of superposition and picks a single state. It's like it knows you're watching and wants to keep its quantum secrets safe.

This paradox raises a cosmic question: Does reality depend on our observation? In the classical world, you'd think that things exist whether you're looking or not. Your pet space cat is either in the box or not, regardless of whether you peek inside. But in the quantum world, particles seem to have this strange dependency on observation.

This phenomenon is what physicists like to call the "measurement problem." It's like the universe is playing a cosmic game of peekaboo with us. It's there, then it's not, then it is again - all depending on whether we're looking or not.

But here's where it gets even weirder. Quantum Quark introduces us to something called "entangled particles." Remember those particles that are quantum BFFs and communicate faster than the speed of light, no matter the distance between them? Well, they add a whole new layer to the paradox.

Imagine you have a pair of entangled particles, and you separate them by cosmic distances. Now, let's say you decide to measure one of them. The moment you do, its quantum buddy - even if it's on the other side of the universe - instantly snaps into a state that's correlated with the one you measured. It's like they're cosmic pen pals who share secrets at the speed of thought.

So, the paradox deepens. Does this mean that the act of observing one particle affects its entangled partner instantaneously, no matter how far apart they are? It's as if the universe is bending the rules of space and time just to keep the quantum party interesting.

Now, you might be wondering, "Do physicists have an answer to this paradox?" Well, it's one of the biggest mysteries in quantum physics, and even the most brilliant minds are still debating it. Some say that maybe there's a hidden layer of reality we haven't discovered yet, while others argue that our observations somehow influence the quantum world.

But one thing is for sure: the Quantum Quark Paradox reminds us that the quantum world is a place of wonder and cosmic head-scratchers. It challenges our very understanding of reality and leaves us with more questions than answers.

So, what's the big takeaway from the Quantum Quark Paradox? It's that the quantum world is full of surprises and paradoxes that keep physicists and space adventurers alike on their toes. It reminds us that the universe is far more mysterious and complex than we can ever imagine.

As we journey deeper into the quantum cosmos in the chapters ahead, keep in mind that Quantum Quark is our trusty guide through this quantum maze. Together, we'll continue to unravel the quantum mysteries that make the universe tick. So, space adventurers, fasten your seatbelts because the quantum adventure is far from over, and the Quantum Quark Paradox is just one cosmic puzzle in this quantum playground.

Chapter 7: Quantum Quark's Time-Traveling Adventure

Hey, intrepid explorers of the cosmos! Welcome back to our quantum journey with Quantum Quark. So far, we've uncovered superpositions, quantum friends, and even danced with paradoxes. But hold onto your space helmets because Chapter 7 is all about "Quantum Quark's Time-Traveling Adventure."

Picture this: Quantum Quark is feeling extra adventurous today, and it's ready to take us on a journey through the fabric of time itself. Yep, you heard that right. We're about to explore one of the most mind-boggling concepts in the quantum universe - time travel.

Now, time travel isn't just a sci-fi fantasy; it's a real concept in the world of quantum physics. It all starts with something called "time dilation." Imagine you're on a super-fast spaceship, zipping through the cosmos at the speed of light. According to Einstein's theory of relativity, time doesn't pass for you the same way it does for someone watching from Earth.

So, while you're having a cosmic adventure, exploring distant galaxies, and chatting with alien civilizations, only a short time might pass for you. But when you return to Earth, you might find that centuries have gone by for the folks back home. It's like you've experienced a bit of time travel, leaping into the future.

But Quantum Quark takes it a step further. It introduces us to the weird and wonderful world of "quantum entanglement" - the same phenomenon we met in previous chapters, where particles become inseparable buddies, no matter the distance between them.

Now, let's say you have two entangled particles, and you separate them by cosmic distances. One particle stays on Earth, and the other goes on a cosmic journey through space at a near-light-speed adventure. Thanks to time dilation, the traveling particle ages much slower than its twin on Earth.

Here's where the time-travel magic happens. When you bring the traveling particle back to Earth and reunite it with its twin, you'd find that, from the perspective of the traveling particle, much less time has passed. It's like they've experienced their very own time-travel adventure into the future.

But wait, there's more! Quantum Quark whispers another secret in your ear. It says that particles don't just experience time differently; they can also communicate across time. Imagine you have an entangled particle pair, and you change the state of one of them in the past. Instantly, the other particle, which might be light-years away, changes its state too, as if it received a message from the past.

This concept is what we call "quantum communication through time." It's like sending a message to your future self or receiving a message from your past self. It's as if the universe has its own cosmic time phone line, connecting particles across the ages.

Now, you might be wondering, "Can we humans use this quantum time-travel trick to build our very own time machines?" Well, not so fast. While these ideas are fascinating, building a practical time machine remains a challenge that even the brightest scientists are still figuring out.

But what this concept shows us is that the quantum world is a place of infinite possibilities, where time isn't always the steady river we imagine. It can warp and bend, and particles can experience it differently, all thanks to the strange and wondrous rules of quantum physics.

So, what's the big takeaway from Quantum Quark's Time-Traveling Adventure? It's that the quantum world is full of surprises, including the possibility of time travel, even if it's just for particles at the moment. It reminds us that our understanding of time is still evolving, and the universe has more mysteries up its cosmic sleeve than we can imagine.

As we venture deeper into the quantum cosmos in the chapters ahead, remember that Quantum Quark is our guide through this time-bending adventure. Together, we'll continue to explore the quantum universe, where even time itself can be a cosmic playground of endless exploration. So, fellow space adventurers, fasten your seatbelts because the quantum adventure is far from over, and who knows what time-traveling secrets the quantum cosmos holds for us next.

Chapter 8: Quantum Quark and the Quantum Twins

Hey there, fellow quantum enthusiasts! Ready for the next leg of our mind-bending journey with Quantum Quark? Well, Chapter 8 is about to introduce us to some cosmic siblings - "Quantum Quark and the Quantum Twins."

So, picture this: Quantum Quark is feeling extra sociable today and decides to take us on a cosmic playdate with two special particles that are like cosmic twins - electrons and positrons. These particles are almost identical, but they have a supercharged twist.

First, meet the electron. You might have heard of this little guy before. It's like the superstar of the atomic world, whizzing around the nucleus in its tiny cosmic orbit. It's got a negative charge, and it's responsible for making all the chemistry in the universe happen. Without electrons, there'd be no atoms, and without atoms, well, the universe would be a pretty boring place.

Now, let's talk about the electron's quantum twin - the positron. Think of it as the electron's mirror image, but with a positive charge instead of a negative one. It's like they're cosmic opposites, yet they have something truly extraordinary in common - they annihilate each other when they meet.

Imagine electrons and positrons as two sides of a cosmic coin. When they collide, they release a burst of energy so intense that it can create some spectacular cosmic fireworks. This process is what we call "annihilation," and it's like a mini Big Bang happening right before our eyes.

Now, here's where things get even more exciting. Quantum Quark tells us about something called "pair production." It's like a cosmic magic trick. Imagine a high-energy photon - a tiny packet of light - zooming through space. When this photon encounters a strong electromagnetic field, it can transform into an electron-positron pair.

Picture this: a photon arrives, and poof it turns into an electron and a positron, like a cosmic two-for-one deal. But here's the kicker: these particles aren't just any particles; they're entangled particles, which means they're like quantum pen pals, even when separated by vast cosmic distances.

Now, let's say we take these entangled electron-positron twins and send them to opposite ends of the universe. When you measure the spin of one particle, you instantly know the spin of the other, no matter how far apart they are. It's like they're cosmic twins with a telepathic connection, communicating faster than the speed of light.

This phenomenon is what we call "entanglement swapping." It's as if the universe has its own cosmic chatroom where particles share their quantum secrets, no matter where they are in the cosmos.

But wait, there's more! Quantum Quark loves to keep us on our toes. It introduces us to a cosmic puzzle known as the "Quantum Zeno Effect." Imagine you have a radioactive atom, and it has a certain chance of decaying over time. In the classical world, you'd expect it to decay eventually, right?

Well, in the quantum world, things get a bit quirky. When you observe the atom frequently, like constantly checking in on it, you can actually slow down its decay. It's like the atom is saying, "I can't decay when you're watching me!" This phenomenon is a playful twist on our understanding of how particles behave and change over time.

So, what's the big takeaway from Quantum Quark and the Quantum Twins? It's that the quantum universe is full of cosmic pairs, from electrons and positrons to entangled particles, each with its own set of mind-blowing tricks. These particles remind us that even in the tiniest corners of the cosmos, there's a universe of wonder waiting to be discovered.

As we venture further into the quantum cosmos in the chapters ahead, keep in mind that Quantum Quark is our guide through this quantum twin adventure. Together, we'll continue to explore the quantum universe, where particles can be both ordinary and extraordinary, depending on how you look at them. So, fellow space adventurers, fasten your seatbelts because the quantum adventure is far from over, and there are always more quantum mysteries to uncover!

Chapter 9: Quantum Quark's Quantum Playground

Hey there, cosmic thrill-seekers! Ready to dive back into the quantum universe with Quantum Quark? Well, in Chapter 9, we're about to step into Quantum Quark's Quantum Playground, where the quantum fun never stops.

Picture this: Quantum Quark is the ultimate cosmic playdate buddy, and it's taking us on a whirlwind tour of the quirkiest and most fascinating attractions in the quantum world.

First up, it's the Quantum Carousel, where particles like to take a wild spin in a game of quantum roulette. Imagine you have an electron, and you're trying to figure out its properties, like its spin. Well, the quantum world says, "Hold my cosmic soda," and spins it in all possible directions at once. It's like a never-ending cosmic dance party, and the electron is the star of the show.

Next, we hop on the Quantum Coaster, where particles can zoom from one energy level to another faster than you can say "quantum leap." Remember those electrons we talked about earlier, the ones that make atomic orbits? Well, they're like thrill-seekers on this coaster, leaping from one energy level to another and releasing quanta of energy like they're on a cosmic rollercoaster ride.

But hold onto your space helmets because Quantum Quark introduces us to the Quantum Mirror Maze, where particles have an identity crisis. Imagine you have two electrons, and you try to swap them. In the classical world, you'd expect them to switch places like obedient particles, right? Not in the quantum world! Particles can be a bit mischievous and refuse to cooperate. They swap, but they also don't swap, all at the same time. It's like they're playing hide-and-seek in a cosmic hall of mirrors.

Now, let's talk about the Quantum Dance Floor, where particles groove to their own quantum beats. Quantum Quark shows us that particles can perform a dance called "quantum superposition." It's like they're doing the cha-cha-cha of the quantum world. They exist in multiple states simultaneously, like a cosmic dance ensemble performing in perfect harmony.

But what about the Quantum Tunnel of Love? In this attraction, particles can magically teleport from one place to another, defying the classical rules of space and time. Quantum tunneling is like the universe's way of saying, "Hey, forget the long road; I'll just take the shortcut." It's responsible for everything from the operation of transistors in your gadgets to nuclear fusion in stars.

As if that's not enough, Quantum Quark leads us to the Quantum Ferris Wheel, where particles can become entangled, forming cosmic connections that boggle the mind. These entangled particles are like cosmic pen pals, and they can communicate faster than the speed of light, no matter the distance between them. It's like having a universal chatroom where particles share secrets at the speed of thought.

But there's one more ride in this quantum playground that'll leave your head spinning - the Quantum Roller Skating Rink. Here, particles can be in multiple places at once, thanks to a phenomenon called superposition. It's like they're gliding through space on invisible roller skates, existing in a cosmic dance of possibilities until you observe them and they decide to settle down.

So, what's the big takeaway from Quantum Quark's Quantum Playground? It's that the quantum world is a vibrant and ever-changing place, where particles like to have a good time and bend the rules of reality. It's like a cosmic amusement park where the attractions are mind-blowing experiments and the rides are the fundamental particles of the universe.

As we venture deeper into the quantum cosmos in the chapters ahead, remember that Quantum Quark is our guide through this quantum wonderland. Together, we'll continue to explore the quantum universe, where particles are the ultimate cosmic playmates, always ready to surprise us with their quantum tricks. So, fellow space adventurers, fasten your seatbelts and get ready for more quantum fun because the quantum adventure is far from over, and there's always something new to discover in Quantum Quark's Quantum Playground.

Chapter 10: The Quantum Quark Enigma

Hey there, cosmic wonder-seekers! Welcome back to our thrilling quantum adventure with Quantum Quark. We've explored quantum playgrounds, danced with particles, and even traveled through time. But now, in Chapter 10, we're about to dive deep into the most enigmatic mystery of all - "The Quantum Quark Enigma."

Imagine this: Quantum Quark is like our cosmic detective, and it's ready to unravel the quantum mysteries that continue to baffle even the brightest minds in the universe.

Now, you might be thinking, "Haven't we already encountered some pretty mind-boggling stuff?" Well, you're absolutely right! We've danced with particles that are in multiple places at once, explored entangled connections across cosmic distances, and even ventured into the realm of time travel. But here's the kicker: the quantum world still holds secrets that defy our understanding.

Quantum Quark introduces us to a concept known as "quantum indeterminacy." It's like the universe's way of saying, "Hey, I like to keep some things to myself." In the classical world, you'd expect that if you knew all the properties of a particle, you could predict its behavior with absolute certainty. But not in the quantum world!

In the quantum universe, there's an inherent uncertainty that makes predicting a particle's exact state impossible. It's like trying to catch stardust with a butterfly net - you'll always have a bit of uncertainty. This fundamental limit on our knowledge of particles is what we call "Heisenberg's Uncertainty Principle," named after physicist Werner Heisenberg.

Heisenberg's Uncertainty Principle tells us that the more precisely we know a particle's position, the less precisely we can know its momentum (which is like its speed and direction) and vice versa. It's as if the universe has a built-in quantum fog that keeps some of its secrets hidden.

But that's not all! Quantum Quark also introduces us to another cosmic puzzle - the "quantum measurement problem." Imagine you're trying to measure a particle's property, like its spin. Well, the act of measuring it can change its state, as if the particle knows it's being watched and decides to behave differently.

This phenomenon challenges our very understanding of what it means to measure something in the quantum world. It's like the universe has a playful side, where particles have a cosmic sense of humor. They dance to their own quantum tune and leave us scratching our heads.

Now, let's talk about something truly mysterious - "quantum decoherence." It's like the universe's way of saying, "I'll keep you guessing." Imagine you have an entangled particle pair, and they're happily communicating across the cosmos. But as soon as you try to observe one of them, it loses its quantum connection with its twin. It's like the universe is saying, "Nope, you can't eavesdrop on our quantum chat."

This concept is a bit like a cosmic magician's trick. The moment you try to peek behind the quantum curtain, the magic disappears, leaving you wondering if it was all just a cosmic illusion.

So, you might be wondering, "Do physicists have any answers to these enigmas?" Well, here's the twist - these mysteries are at the heart of the quantum world, and even the brightest minds are still exploring them. Some physicists suggest that the quantum world may be inherently uncertain, and our measurements may always carry a bit of mystery.

But here's what makes the Quantum Quark Enigma so captivating - it reminds us that the quantum universe is a place of wonder, where the rules of reality are ever-changing and often mind-bending. It challenges us to question our understanding of the cosmos and keeps us on our toes as we explore its quantum secrets.

So, what's the big takeaway from the Quantum Quark Enigma? It's that the quantum world is a place of eternal fascination and cosmic mystery. It invites us to embrace the unknown and reminds us that there are still enigmas waiting to be unraveled in the farthest corners of the universe.

As we venture further into the quantum cosmos in the chapters ahead, remember that Quantum Quark is our trusty guide through this quantum maze. Together, we'll continue to explore the quantum universe, where even the most perplexing enigmas are invitations to dive deeper into the cosmic unknown. So, fellow space adventurers, fasten your seatbelts because the quantum adventure is far from over, and the Quantum Quark Enigma is just one more cosmic riddle waiting to be solved.

Chapter 11: Quantum Quark's Quantum Quest

Hey there, cosmic explorers! We've journeyed through quantum playgrounds, danced with particles, and unraveled enigmas that boggle the mind. But in Chapter 11, it's time to embark on Quantum Quark's Quantum Quest - a thrilling cosmic adventure into the heart of the quantum universe.

Imagine this: Quantum Quark is our trusty guide, and it's ready to lead us on a quest to uncover the hidden treasures of the quantum cosmos. It's like a cosmic treasure map, and each discovery is a sparkling gem waiting to be found.

Our first stop on this quantum quest is the Land of Quantum Superpositions. Picture a place where particles can exist in multiple states at once, like a cosmic multitasker juggling a dozen things. It's a world where a particle can be here and there, up and down, all at the same time. This superposition dance is at the core of quantum weirdness, and it's our first clue on this quest.

As we journey deeper into the quantum cosmos, Quantum Quark introduces us to the Quantum Entanglement Oasis. It's a serene place where particles become inseparable quantum buddies, no matter the cosmic distance between them. This entangled connection is like a secret bond that defies the rules of space and time, and it's our second clue on this quest.

Next up, we step into the Quantum Time-Traveling Tunnel. It's a cosmic gateway that allows particles to leap through time, experiencing the past, present, and future all at once. This time-bending phenomenon is like a portal to the mysteries of the universe, and it's our third clue on this quest.

But the quantum universe isn't done surprising us. Quantum Quark takes us to the Quantum Maze of Uncertainty. It's a labyrinth of unpredictability, where particles like to keep some of their secrets hidden. This maze challenges our understanding of what it means to know something in the quantum world, and it's our fourth clue on this quest.

Now, let's venture into the Quantum Wonderland of Paradoxes. It's a place where the rules of reality seem to twist and turn, leaving us with more questions than answers. We've danced with particles that change when we observe them, and we've encountered mysteries that defy logic. This wonderland is our fifth clue on this quest.

As we journey through these quantum realms, we start to piece together a grand cosmic puzzle. It's like assembling a jigsaw puzzle of the universe, where each quantum discovery is a piece that fits into the larger picture.

But here's the twist - the quantum quest isn't just about finding answers; it's about embracing the wonder of the unknown. Quantum Quark reminds us that the quantum universe is a place of eternal fascination and mystery, where every question leads to more questions, and every discovery reveals more wonders.

So, what's the big takeaway from Quantum Quark's Quantum Quest? It's that the journey through the quantum cosmos is an adventure of a lifetime, where the destination is not as important as the discoveries along the way. It's a reminder that the universe is vast, complex, and ever-changing, and our quest to understand it is a quest that never ends.

As we continue to explore the quantum universe in the chapters ahead, remember that Quantum Quark is our cosmic companion on this quest. Together, we'll delve deeper into the quantum mysteries that make the universe tick, and we'll embrace the thrill of the unknown as we venture into the quantum cosmos. So, fellow space adventurers, fasten your seatbelts because the quantum adventure is far from over, and Quantum Quark's Quantum Quest is a journey that leads to the stars and beyond!

Chapter 12: Quantum Quark and the Quantum Conundrum

Hey there, fellow quantum explorers! Ready to dive back into the cosmic rabbit hole with Quantum Quark? Well, in Chapter 12, we're about to tackle one of the most puzzling challenges of the quantum universe - "Quantum Quark and the Quantum Conundrum."

Picture this: Quantum Quark is like our trusty cosmic detective, and it's got its quantum magnifying glass out, ready to solve the mysteries of the quantum cosmos. But this conundrum? It's a doozy.

Let's start with a quick recap. In our quantum journey, we've encountered particles that can be in multiple places at once, entangled particles that communicate faster than the speed of light, and even particles that can leap through time. But here's where it gets mind-bendingly interesting.

Quantum Quark introduces us to the concept of "quantum teleportation." No, it's not like the teleporters you see in sci-fi shows, where you step in one end and pop out the other. Quantum teleportation is more like a cosmic fax machine for particles.

Imagine you have Particle A and Particle B. They're entangled, which means they have a deep quantum connection. Now, let's say you want to send Particle A's information to a friend on the other side of the universe, where Particle B is waiting. It seems like a cosmic FedEx job, right?

Well, not in the quantum world! Quantum teleportation lets you transmit Particle A's information to Particle B instantaneously, no matter how far apart they are. It's like you're sending Particle A's quantum essence across the cosmos, and Particle B becomes an exact copy of Particle A. It's a bit like magic, but it's a real phenomenon in the quantum universe.

But here's the twist - in the process of quantum teleportation, you have to "destroy" Particle A. It's not like you're disintegrating it, but you're measuring its quantum state so precisely that it can no longer exist as it was. So, Particle A sacrifices itself to send its quantum information to Particle B. It's like a cosmic game of quantum tag, where one particle "tags" the other and becomes something new in the process.

Now, this quantum conundrum raises a cosmic question: Is teleportation just a fancy way of moving information, or is it moving something more profound, like the essence of the particle itself? It challenges our understanding of what particles are and how they exist in the quantum world.

But wait, there's more! Quantum Quark loves to keep us on our toes. It introduces us to the "many-worlds interpretation." This theory suggests that every quantum event creates multiple branching universes, each representing a different outcome.

Imagine you're faced with a quantum choice, like the spin of an electron. According to the many-worlds interpretation, the universe splits into multiple versions - one for each possible outcome. In one universe, the electron's spin is up; in another, it's down. It's like the cosmic multiverse is constantly branching and creating new realities.

This interpretation raises a cosmic conundrum of its own. If every quantum choice spawns a new universe, then there are an infinite number of parallel universes out there, each with its own version of reality. It's like the universe is a cosmic Choose Your Own Adventure book with countless pages.

So, what's the big takeaway from Quantum Quark and the Quantum Conundrum? It's that the quantum universe is a place of endless possibilities and mind-bending mysteries. Teleportation challenges our understanding of particles and their essence, while the many-worlds interpretation opens the door to a multiverse of realities.

As we venture deeper into the quantum cosmos in the chapters ahead, remember that Quantum Quark is our guide through this quantum maze. Together, we'll continue to explore the quantum universe, where every conundrum is an invitation to push the boundaries of our cosmic curiosity. So, fellow space adventurers, fasten your seatbelts because the quantum adventure is far from over, and the Quantum Conundrum is just one more cosmic puzzle waiting to be solved!

Chapter 13: Quantum Quark's Quantum Mechanics

Hey there, cosmic curious minds! Welcome back to the quantum odyssey with our trusty guide, Quantum Quark. In Chapter 13, we're diving headfirst into the fascinating world of "Quantum Quark's Quantum Mechanics." It's like peering through a cosmic magnifying glass to uncover the inner workings of the quantum universe.

Now, if you've been with us on this quantum journey, you know we've explored some mind-bending stuff - particles that teleport, entangled buddies that chat across the cosmos, and even the mysterious quantum fog of uncertainty. But here's where it all comes together, in the realm of quantum mechanics.

So, what's quantum mechanics, you ask? Think of it as the rulebook, the instruction manual, the cosmic code that governs the behavior of the tiniest things in the universe - particles like electrons, photons, and even quantum-sized dust bunnies. It's like the quantum universe's way of saying, "This is how we roll!"

One of the biggest stars of quantum mechanics is something you might have heard of - the "wave-particle duality." It's like particles have a cosmic identity crisis. Sometimes they act like solid little marbles, bouncing around and doing their thing. Other times, they behave like waves, spreading out and creating interference patterns. It's like they're playing both hide-and-seek and musical chairs at the same time.

Now, you might be wondering, "How can something be both a particle and a wave?" Well, that's the quantum quirkiness of it all. In the quantum world, particles like to keep us guessing, and wave-particle duality is their way of saying, "We're not just one thing; we're everything."

But hold on, the quantum fun doesn't stop there! Quantum Quark takes us to another star attraction - "superposition." It's like particles are doing a cosmic juggling act, existing in multiple states at once. Imagine Schrödinger's famous cat, which is both alive and dead until you open the box and check. It's like the universe's way of saying, "Don't put your eggs in one cosmic basket; we like to keep our options open."

And then there's "entanglement." This cosmic phenomenon is like particles forming a quantum buddy system, where they become linked in ways that boggle the mind. Change the state of one particle, and its entangled twin instantly changes too, no matter how far apart they are. It's like they have a cosmic hotline where they share their secrets faster than a quantum text message.

But here's the twist - quantum mechanics isn't just a collection of strange phenomena; it's also a set of mathematical tools that help us make sense of the quantum world. Think of it as the quantum toolbox, where scientists use equations and formulas to describe the behavior of particles.

One of the quantum mechanics all-stars is the Schrödinger equation. It's like the quantum universe's favorite song, the one that tells particles how to dance through space and time. It helps us predict where particles are likely to be and what they're likely to do.

And then there's Heisenberg's Uncertainty Principle, which we've met before. It's like the quantum referee blowing the whistle and saying, "You can't know everything about a particle's position and momentum at the same time." It sets a cosmic limit on what we can measure in the quantum world.

Now, I know what you might be thinking - "This all sounds like a lot of cosmic complexity." And you're absolutely right! Quantum mechanics is like the symphony of the cosmos, where particles play their own unique notes, creating a cosmic masterpiece.

But here's the beauty of it all - quantum mechanics isn't just a set of rules; it's an invitation to explore the quantum universe's deepest secrets. It challenges us to push the boundaries of our understanding and embrace the wonder of the unknown.

So, what's the big takeaway from Quantum Quark's Quantum Mechanics? It's that the quantum universe is a place of infinite possibilities and cosmic complexity. Quantum mechanics is like the universal language that lets us decode the quantum riddles and dance to the quantum tune.

As we venture further into the quantum cosmos in the chapters ahead, remember that Quantum Quark is our cosmic mentor through this quantum journey. Together, we'll continue to explore the quantum universe, where particles are the cosmic performers in the grandest show of all. So, fellow space adventurers, fasten your seatbelts because the quantum adventure is far from over, and Quantum Quark's Quantum Mechanics is our backstage pass to the quantum stage of the universe!

Greetings, fearless quantum explorers! We've journeyed through the quantum universe, danced with particles, and unraveled enigmas that boggle the mind. But in Chapter 14, it's time to face Quantum Quark's Quantum Challenge - a cosmic adventure that will test the limits of our quantum understanding.

Imagine this: Quantum Quark is like a cosmic coach, and it's sending us on a quest to tackle some of the most perplexing challenges in the quantum world. It's like a cosmic scavenger hunt where each challenge is a riddle waiting to be solved.

Our first challenge takes us to the Quantum Labyrinth of Superposition. Picture a maze where particles are in multiple states at once, like a quantum multitasker juggling a dozen things. Your mission? Navigate this maze and find the exit while particles zigzag around you, existing everywhere and nowhere simultaneously.

Next up, we face the Quantum Entanglement Escape Room. You're locked in a room with a pair of entangled particles, and you need to decipher their quantum connection to unlock the door. It's like a cosmic puzzle where their states are intertwined, and you must figure out how to break free.

But the quantum universe isn't done challenging us. Quantum Quark leads us to the Quantum Teleportation Time Trial. Imagine you have to teleport a particle's information across the cosmos while preserving its quantum essence. It's a race against the cosmic clock, and you must complete the teleportation before the universe blinks.

Our next challenge is the Quantum Uncertainty Escape. You're in a room filled with particles, and you must measure their positions and momenta without violating Heisenberg's Uncertainty Principle. It's like a quantum obstacle course where the rules keep changing, and you must tread lightly to succeed.

Chapter 14: Quantum Quark's Quantum Challenge

But wait, there's more! Quantum Quark introduces the Quantum Parallel Universe Puzzle. You're presented with a quantum choice, like the spin of an electron. Your mission? Create a branching universe for every possible outcome, just like the many-worlds interpretation suggests. It's like juggling a cosmic circus of parallel realities.

As we take on these quantum challenges, we start to piece together a cosmic puzzle. It's like solving a quantum Rubik's Cube, where each challenge is a piece that fits into the larger picture.

But here's the twist - the quantum challenge isn't just about finding solutions; it's about embracing the thrill of the unknown. Quantum Quark reminds us that the quantum universe is a place of endless wonder, where every challenge is an invitation to push the boundaries of our cosmic curiosity.

So, what's the big takeaway from Quantum Quark's Quantum Challenge? It's that the journey through the quantum cosmos is an adventure of a lifetime, where the challenges are cosmic riddles waiting to be unraveled. It's a reminder that the universe is vast, complex, and ever-changing, and our quest to understand it is a quest that never ends.

As we venture deeper into the quantum cosmos in the chapters ahead, remember that Quantum Quark is our cosmic mentor on this journey. Together, we'll tackle the quantum challenges that make the universe tick, and we'll embrace the thrill of the unknown as we venture into the quantum frontier. So, fellow space adventurers, fasten your seatbelts because the quantum adventure is far from over, and Quantum Quark's Quantum Challenge is a cosmic test of our quantum mettle.

Chapter 15: Quantum Quark's Quantum Celebration

Hey there, fellow cosmic travelers! Can you believe it? We've journeyed through the quantum cosmos with our trusty guide, Quantum Quark, and now, in Chapter 15, it's time for a grand celebration of all thing's quantum. It's like the ultimate cosmic party where particles are the stars of the show, and mysteries are the confetti in the quantum breeze.

Imagine this: Quantum Quark is throwing a quantum bash to celebrate our incredible journey through the quantum universe. It's like a cosmic carnival with all the quantum wonders we've encountered along the way, from particles that teleport to entangled buddies that communicate faster than the speed of light.

Our first stop in this quantum celebration is the Quantum Superposition Spectacular. Picture particles existing in multiple states at once, like quantum acrobats performing a mind-bending show. It's like watching a cosmic magic act where particles defy the rules of reality and leave us in awe.

Next up, we have the Quantum Entanglement Extravaganza. This is where particles become quantum dance partners, sharing secrets across the vast cosmic stage. It's like a never-ending waltz where entangled particles perform a mesmerizing ballet, showing us the beauty of quantum connections.

But the party doesn't stop there! Quantum Quark leads us to the Quantum Teleportation Tango. Imagine particles transmitting their quantum essence across the universe in a dazzling dance of information. It's like a cosmic ballet where particles twirl and teleport in perfect harmony, leaving us breathless.

Our next celebration is the Quantum Uncertainty Carnival. Here, particles play cosmic games of hide-and-seek, challenging us to uncover their positions and momenta without violating Heisenberg's Uncertainty Principle. It's like a quantum funhouse where the rules keep changing, and surprises are around every corner.

But wait, there's more! Quantum Quark unveils the Quantum Multiverse Masquerade. Picture countless parallel universes, each with its own version of reality, all dancing to the quantum beat. It's like a cosmic ball where alternate realities don masks and twirl to the music of the multiverse.

As we revel in the quantum celebration, we realize that the quantum universe is not just a place of strange phenomena; it's a testament to the wonders of the unknown. Quantum Quark reminds us that every challenge, every mystery, and every cosmic conundrum is an invitation to explore the boundless frontiers of the universe.

So, what's the big takeaway from Quantum Quark's Quantum Celebration? It's that the quantum universe is a place of infinite possibilities and cosmic complexity, and our journey through it has been nothing short of extraordinary. It's a reminder that the universe is a vast, ever-changing tapestry, and our exploration of it is a cosmic dance of curiosity and wonder.

As we bid farewell to the quantum celebration and our journey through the quantum cosmos, let's remember that Quantum Quark is our cosmic companion, always ready to guide us through the mysteries of the universe. Together, we've uncovered the quantum secrets that make the universe tick, and we've embraced the thrill of the unknown as we ventured into the quantum frontier.

So, fellow space adventurers, as we conclude this cosmic dance, let's raise a quantum toast to the wonders of the universe, to the mysteries that keep us curious, and to the endless possibilities that await us in the vast quantum cosmos. The quantum adventure may be over for now, but the cosmic curiosity it has ignited will continue to shine brightly in our hearts. Cheers to Quantum Quark's Quantum Celebration, and here's to the next cosmic journey that awaits us among the stars!

About The Author

Welcome to the world of Tihirou Nicol, a multifaceted individual whose life journey is as diverse as it is inspiring. From a successful civil engineering career to entrepreneurial ventures and an unshakable passion for soccer, let's delve into the story of this remarkable author.

Background and Education

Tihirou Nicol's journey began in the vibrant city of Freetown, Sierra Leone. He found his early love for soccer while attending the esteemed Methodist Boys High School, a place where his passion for the sport first ignited. His academic pursuits then led him to Fourah Bay College, one of Africa's oldest universities, where he embarked on a journey to attain a Bachelor's Degree in Civil Engineering. Driven by a relentless thirst for knowledge, he continued his educational journey, achieving a Master's Degree in Civil Engineering.

A Journey of Transformation

Tihirou professional path initially specialized in the complex world of road construction. Yet, life is a journey of transformation, and his took an intriguing turn, leading him to explore a myriad of ventures beyond his engineering roots.

From Civil Engineering to Business

Following a fulfilling career with Kier Construction in the United Kingdom, Tihirou made a bold leap into the realm of entrepreneurship. Undaunted by the challenges, he embarked on an entrepreneurial adventure, founding his own property business in London before extending his ventures to the United States.

Embracing My Passion: Soccer

Soccer isn't just a sport for Tihirou; it's a lifelong passion. From his formative years, he actively participated in the beautiful game until an injury altered his course. Nonetheless, his love for soccer remained unwavering, leading him to become the proud owner of a soccer shopping outlet, where he continues to contribute to the sport that has left an indelible mark on his life.

A Commitment to Community

While his entrepreneurial endeavors keep him busy, Tihirou finds solace in giving back to the community. He extends active support to his local soccer team, East End Lions, and champions the legendary Liverpool Football Club, both holding a special place in his heart.

Motivated by Dreams

Throughout his journey, visionaries like Sir Richard Branson have been a wellspring of inspiration for Tihirou. His aspiration is clear: to make a positive impact on people's lives, helping them achieve their dreams and desires.

Philosophy and Approach

Hard work, perseverance, and creativity form the bedrock of Tihirou 's approach to life and business. He's a relentless seeker of innovative solutions to surmount challenges, ever in pursuit of success.

A Future Full of Ambitions

Looking ahead, Tihirou envisions a future marked by the expansion of shopping outlets across Sierra Leone, London, and Maryland, USA. These endeavors will not only contribute to economic growth but also reflect his unwavering passion for soccer.

Thank you for embarking on this journey of exploration and growth with Tihirou Nicol. Together, let's build a world where dreams flourish and the universal love for soccer brings people from diverse backgrounds together.

www.ingramcontent.com/pod-product-compliance
Lightning Source LLC
Chambersburg PA
CBHW082228290526

45794CB00009B/3719